WHY DON'T YOU
GET OFF
YOUR
PHONE
AND LEARN
SOMETHING
NEW INSTEAD?

WHY DON'T YOU GET OFF YOUR PHONE AND LEARN SOMETHING
NEW INSTEAD?

An Hachette UK Company
www.hachette.co.uk

Summersdale Publishers Ltd
Part of Octopus Publishing Group Limited
Carmelite House
50 Victoria Embankment
LONDON
EC4Y 0DZ
UK

www.summersdale.com

Printed and bound in Croatia

ISBN: 978-1-78685-282-3

Substantial discounts on bulk quantities of Summersdale books are available to corporations, professional associations and other organisations. For details contact general enquiries: telephone: +44 (0) 1243 771107 or email: enquiries@summersdale.com.

Disclaimer
Neither the author nor the publisher can be held responsible for any loss or claim arising out of the use, or misuse, of the suggestions made herein.
All information included in this book is correct at the time of going to press.

WHY DON'T YOU

GET OFF YOUR PHONE

AND LEARN SOMETHING NEW INSTEAD?

KATE FREEMAN

summersdale

CONTENTS

INTRODUCTION

We're all guilty of reaching for our phones many more times a day than is necessary – whether it's to settle a debate with a search engine, browse social media or simply peruse its contents out of boredom. In moderation, this is fine, but too much screen time is not healthy.

So why not set yourself the challenge to use your phone less? You could leave it in another room when you go to bed, so you're not tempted to reach for it as soon as you wake up. Or switch it off when you're engaging in other activities or spending time with friends. You might also find that something as simple as tweaking the settings to offer fewer notifications has you reaching for your device less often.

To help you along the way, this book is here as your handy guide. Animate your interest and stimulate your senses with the host of trivia, activities, experiments, projects and practical tips in the following pages. Rest those screen-weary eyes and learn something new instead!

NATURE

Whether you've always wanted to teach a parrot to talk, outrun a crocodile or watch a meteor shower, now is the time to get back to nature.

ATTRACT BUTTERFLIES TO YOUR GARDEN

Butterflies are in decline, so encourage them into your outdoor space. The best way to do this is to present them with flowers in your borders or in pots and hanging baskets. Adult butterflies in general are partial to reds, yellows, oranges, pinks and purples, so gear your garden towards their floral colour preferences. Flowers with tubes are also a plus as they often have the best sources of nectar. Try to stagger the bloom cycles in your garden to attract the winged wonders all year round.

WORK OUT WHEN THE SUN IS GOING TO SET

In case you find yourself in a sticky situation out in nature without a watch or phone, you might find this skill useful. First, find a high vantage point where you can see both the sun and the horizon. Avoiding looking at the sun directly, hold your hand up in front of you, palm facing towards you and with your fingers together, and count how many times your hand fits between the sun and the horizon. Use your fingers as fractions if it's not an even number of hands. One hand equals an hour and one finger 15 minutes before sunset.

TEACH A PARROT TO TALK

The top three mimics of the bird world are the African grey parrot, Amazon parrots and the oft-overlooked budgie. When trying to master the art of bird chat, be patient – repetition is the name of the game and your feathered friend's responses will grow clearer with practice. Set up camp in a quiet room with low lighting, free from distractions, and repeat a word over and over. Alternatively, record your chosen word and play it on a loop. Each bird is different: some will mimic better than others and it's thought that they are more responsive to higher-pitched voices.

BUILD A BIRD FEEDER

1. Take an empty tin can of any size and use an opener to remove the bottom. Use a piece of thick sandpaper to carefully smooth any rough edges.

2. Take a thick piece of cardboard and trace around one end of the tin can, then cut out the circle. Do this twice.

3. Take the cardboard circles and remove the middle third.

4. Glue a cardboard ring to each end to cover the edges of the can. This will stop the seed from escaping when the can is hanging.

5. Take two wooden rods or sticks at least half the length of the can and glue to the inside, so that a good amount of the stick protrudes. (You might need to make a hole through the cardboard ring to do this.) These will form perches at either end of your feeder.

6. Tie string around the middle of the can, fill it with bird seed and hang horizontally from a tree.

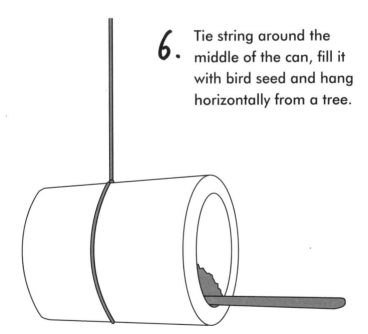

MAKE A NATURAL BIRD FEEDER

Simply take a large, under-ripe pumpkin, hollow it out, fill it with bird food and hang it from a tree; or use a large pine cone, covered with peanut butter and seeds.

WATCH A METEOR SHOWER

For optimum meteor shower viewing you need to get away from artificial light. That means leaving the city and getting away from any built-up areas and into the dark of the countryside. Don't bother with a telescope or binoculars – this massively reduces the amount of sky you can view at any one time and it's all about scanning the great beyond for those elusive shooting stars. Your best bet is the annual Perseid meteor shower, which hits in July/August, and at its peak offers as many as 100 meteors an hour whizzing overhead. Wrap up warm and hope for clear skies!

HOLD A CHICKEN

Many people keep chickens at home, so here is some advice for the next time you visit that bird-keeping friend (or if you get some chickens of your own). Never hold a chicken by its wings. Place its feet side by side and slip the first and second fingers of one hand on either side of them. Keep the chicken's wings folded naturally at its sides and, cupping your hand around the body, use your thumb to cover the primary flight feathers as far as possible. Your other hand can then be placed against the breast to balance the chicken.

GAUGE THE PROXIMITY OF A STORM

During a storm you'll always hear the thunder after the lightning because light travels faster than sound. With the speed of light at 299,792,458 metres per second and the speed of sound 344 metres per second, you can understand why there's quite a delay. Count the seconds between lightning and thunder and multiply by 344 to work out how far away (in metres) the lightning struck. Alternatively, count the number of seconds and divide by five for the number of miles – so 10 seconds between flash and crash would equal 2 miles.

OUTRUN A CROCODILE

While one should never take this up as a
sport or try it on a whim, should you ever
find you need to escape a crocodile this
advice might be useful. Crocodiles are
able to move at speeds of up to 11 mph
on land, but if you are fit and healthy you
should have no problem outrunning the
scaly one. Some sources claim crocodiles
struggle to change direction quickly,
so you could do a bit of zigzagging.
However, the fastest route from A to B
is in a straight line and crocodiles only
experience relatively short bursts of
speed, so do what you have to do.

KNOW YOUR CROCODILES FROM YOUR ALLIGATORS

Alligators sport more rounded snouts than crocodiles, and the fourth tooth in a croc's lower jaw sticks out when its mouth is closed, while a gator's fits snugly into the upper jaw.

KNOW YOUR ELEPHANTS

An African elephant has large tusks and ears, a slightly hollowed back, two bumps at the tip of its trunk and is dark grey; an Asian elephant is lighter grey, smaller, has a slightly arched back and one bump at the tip of its trunk.

LEARN ANIMAL FAMILY NAMES

How is your knowledge of male, female and baby animal names? Brush up here.

ANIMAL	MALE	FEMALE	BABY
Goat	Billy	Nanny	Kid
Deer	Buck	Doe	Fawn
Horse	Stallion	Mare	Foal
Sheep	Ram	Ewe	Lamb
Fox	Dog	Vixen	Cub
Seal	Bull	Cow	Pup
Pig	Boar	Sow	Piglet
Bee	Drone	Queen/Worker	Larva
Ferret	Hob	Jill	Kit
Goose	Gander	Goose	Gosling
Hawk	Tiercel	Hen	Eyas
Kangaroo	Buck	Doe	Joey
Duck	Drake	Duck	Duckling
Panda	Boar	Sow	Cub
Wallaby	Jack	Jill	Joey
Whale	Bull	Cow	Calf

KNOW YOUR COLLECTIVE CREATURE NAMES

Bats
A cauldron

Cobras
A quiver

Cockroaches
An intrusion

Crabs
A consortium

Crows
A murder

Eagles
A convocation

Hippos
A thunder

Jaguars
A shadow

Jellyfish
A bloom

Lapwings
A deceit

Larks
An exaltation

Lemurs
A conspiracy

Magpies
A tiding

Otters
A romp

Owls
A parliament

Parrots
A pandemonium

Rhinos
A crash

Squid
An audience

Tigers
An ambush

Salamanders
A maelstrom

Sharks
A shiver

SCIENCE

Ever held a potato in your hand and wondered
how to get power from it? Well wonder no more.
Here you will also learn how to measure humidity,
grow crystals and build your own volcano.

GET POWER FROM A POTATO

1. Cut a potato in half and cut a small slit into the flat side of each half.

2. Take two pennies and two pieces of copper wire. Wrap a piece of copper wire around each penny a few times.

3. Place a penny into the slit in each potato half.

4. Take another copper wire and wrap it around a zinc-plated nail, then stick the nail into one of the potato halves.

5. Take the wire connected to the penny in the half of potato with the nail and wrap some of it around another nail. Stick that second nail into the other potato half.

6. When you connect the two loose ends of the copper wires to a light bulb or LED it will light up.

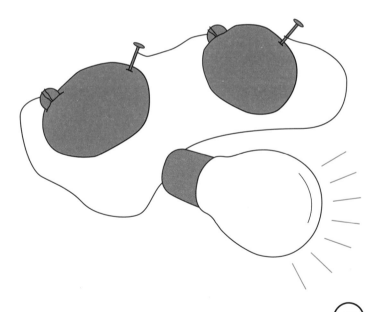

MEASURE HUMIDITY

Human hair gets longer in moist air. So one way of measuring humidity is to fasten a long human hair between two points. If the humidity falls, and the air gets drier, the hair will snap. If it becomes loose then the air has become more humid. The greater the humidity of the air, the more likely it is to rain.

GROW CRYSTALS

Salt, alum and copper sulphate are good crystal-growing substances to experiment with. (Find alum at your local chemist and copper sulphate online.) Take a clean jar, fill it two-thirds full of hot water, add some of your substance and stir until it dissolves. Keep adding and stirring until no more of the substance will dissolve. Tie a piece of string or thread to the middle of a pencil and rest it on the rim of the jar, with the thread dipped into the solution. Leave undisturbed for a week and see the crystals build up on the thread. Add food colouring or ink for more colourful results!

STICK A BALLOON TO THE WALL

This is a good one to impress young children. Take a balloon and rub it against your clothing or hair for 10 seconds to create static, then hold it against a wall and watch how it stays put when you remove your hand.

LEARN THE FIRST 30 ELEMENTS

Hydrogen
H

Helium
He

Lithium
Li

Beryllium
Be

Boron
B

Carbon
C

Nitrogen
N

Oxygen
O

Fluorine
F

Neon
Ne

Sodium
Na

Magnesium
Mg

Aluminium
Al

Silicon
Si

Phosphorus
P

Sulphur
S

Chlorine
Cl

Argon
Ar

Potassium
K

Calcium
Ca

Scandium
Sc

Titanium
Ti

Vanadium
V

Chromium
Cr

Manganese
Mn

Iron
Fe

Cobalt
Co

Nickel
Ni

Copper
Cu

Zinc
Zn

PLAY GLOW-IN-THE-DARK BOWLING

You'll need six glow sticks, six plastic water bottles (labels removed) nearly full of water, a ball such as a basketball or football, and something to keep score. Add a glow stick to each bottle (don't forget to replace the tops), set up your 'bowling pins' in a triangle formation and away you go!

WOW YOUR FRIENDS WITH SPACE FACTS

Did you know that if you could find a body of water big enough to hold it, Saturn would float? Saturn is mostly made up of gas and is therefore less dense than water, which would allow it to float, unlike other planets in our solar system which all have higher densities than water. With an impressive diameter of 74,898 miles (120,536 km) – compared with Earth's 7,926 miles (12,756 km) – Saturn is the second largest planet in our solar system behind Jupiter, which has a diameter of a whopping 88,846 miles (142,984 km).

 # KNOW YOUR BONES

1. Cranium

2. Atlas

3. Mandible

4. Clavicle

5. Scapula

6. Sternum

7. Humerus

8. Vertebrae

9. Radius

10. Ulna

11. Sacrum

12. Coccyx

13. Femur

14. Patella

15. Fibula

16. Tibia

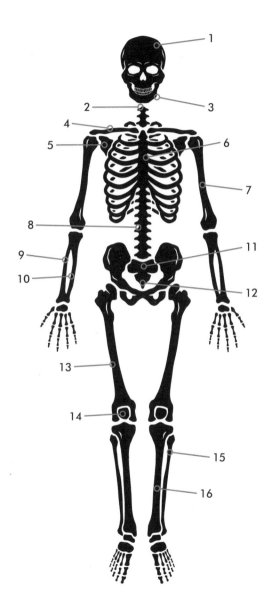

MAKE JELLY
SET QUICKLY

Need a bowl of jelly in a hurry?
Simply add a third less hot water
than stated on the packet and top
up the jelly mixture with ice. This will
bring down the temperature of the
mixture quickly and speed up the
setting process.

CREATE GIANT BUBBLES

1. Make your own bubble mixture with washing-up liquid and water in a washing-up bowl. The mixture should comprise approximately three parts water to one part washing up liquid, but use trial and error to get your mix right for your type of soap and water. Supercharge it by adding a drop or two of glycerine, available from chemists and craft shops.

2. To make the wand you'll need two lengths of dowelling about 30 cm long (or you could use two large wooden spoons), some string and a small weight, like a threaded nut.

3. Cut a 90-cm length of string and thread it through the bolt, and tie the two ends of the string together. Create an upside-down triangle shape with the string with the bolt at the bottom, then attach the top two corners to the ends of your wands with tape.

4. Dip the string into the bubble mixture with the wands together and gently lift out, parting the sticks and waving the string to make your giant bubbles.

MAKE A HOME-MADE VOLCANO

1. Take a large plastic bottle, cut out the middle segment and reattach the top with tape.

2. Using wet strips of newspaper and glue, build up a volcano shape around the bottle.

3. Once dry, paint it an appropriate colour.

4. Now for the eruption – best to move out into the garden for this. Spoon 4 tbsp bicarbonate of soda into the bottle and, in a separate container, add a few drops of red and yellow food colouring to 250 ml of vinegar.

5. Use a funnel to pour the vinegar into the bottle, but remove the funnel quickly as your eruption will follow soon after!

HISTORY

Wow friends, family and anyone who will
listen with your new-found historical knowledge.
From English monarchs and prime ministers
to US presidents, not to mention the
towns and cities of Roman Britain –
it's all here for the learning.

MEMORISE THE ENGLISH MONARCHS

NORMANS

William I
1066–1087

William II
1087–1100

Henry I
1100–1135

Stephen
1135–1154

PLANTAGENETS

Henry II
1154–1189

Richard I
1189–1199

John
1199–1216

Henry III
1216–1272

Edward I
1272–1307

Edward II
1307–1327

Edward III
1327–1377

Richard II
1377–1399

HOUSE OF LANCASTER

Henry IV
1399–1413

Henry V
1413–1422

Henry VI
1422–1461
1470–1471

HOUSE OF YORK

Edward IV
1461–1470
1471–1483

Edward V
1483

Richard III
1483–1485

TUDORS

Henry VII
1485–1509

Henry VIII
1509–1547

Edward VI
1547–1553

Jane Grey
1553

Mary I
1553–1558

Elizabeth I
1558–1603

STUARTS

James I
1603–1625

Charles I
1625–1649

COMMONWEALTH

Oliver
Cromwell
1649–1658

Richard
Cromwell
1658–1659

STUARTS
(restored)

Charles II
1660–1685

James II
1685–1688

William III
1689–1702

Mary II
1689–1694

Anne
1702–1714

HOUSE OF HANOVER

George I
1714–1727

George II
1727–1760

George III
1760–1820

George IV
1820–1830

William IV
1830–1837

Victoria
1837–1901

SAXE-COBURG-GOTHA

Edward VII
1901–1910

WINDSOR

George V
1910–1936

Edward VIII
1936

George VI
1936–1952

Elizabeth II
1952–

BRUSH UP ON YOUR US PRESIDENTS

George Washington
1789–1797

John Adams
1797–1801

Thomas Jefferson
1801–1809

James Madison
1809–1817

James Monroe
1817–1825

John Quincy Adams
1825–1829

Andrew Jackson
1829–1837

Martin Van Buren
1837–1841

William Henry Harrison
1841

John Tyler
1841–1845

James K. Polk
1845–1849

Zachary Taylor
1849–1850

Millard Fillmore
1850–1853

Franklin Pierce
1853–1857

James Buchanan
1857–1861

Abraham Lincoln
1861–1865

Andrew Johnson
1865–1869

Ulysses S. Grant
1869–1877

Rutherford B. Hayes
1877–1881

James A. Garfield
1881

Chester A. Arthur
1881–1885

Grover Cleveland
1885–1889

Benjamin Harrison
1889–1893

Grover Cleveland
1893–1897

William McKinley
1897–1901

Theodore
Roosevelt
1901–1909

William Howard
Taft
1909–1913

Woodrow
Wilson
1913–1921

Warren G.
Harding
1921–1923

Calvin
Coolidge
1923–1929

Herbert
Hoover
1929–1933

Franklin D.
Roosevelt
1933–1945

Harry S.
Truman
1945–1953

Dwight D.
Eisenhower
1953–1961

John F.
Kennedy
1961–1963

Lyndon B.
Johnson
1963–1969

Richard M.
Nixon
1969–1974

Gerald R. Ford
1974–1977

Jimmy Carter
1977–1981

Ronald Reagan
1981–1989

George Bush
1989–1993

Bill Clinton
1993–2001

George W.
Bush
2001–2009

Barack Obama
2009–2017

Donald Trump
2017–

KNOW YOUR TWENTIETH- AND TWENTY-FIRST-CENTURY PRIME MINISTERS

Marquess of
Salisbury
1895–1902

Arthur James
Balfour
1902–1905

Henry Campbell-
Bannerman
1905–1908

H. H. Asquith
1908–1916

David Lloyd
George
1916–1922

Andrew Bonar
Law
1922–1923

Stanley
Baldwin
1923–1924

Ramsay
MacDonald
1924

Stanley
Baldwin
1924–1929

Ramsay
MacDonald
1929–1935

Stanley
Baldwin
1935–1937

Neville
Chamberlain
1937–1940

Winston
Churchill
1940–1945

Clement Attlee
1945–1951

Winston
Churchill
1951–1955

Anthony Eden
1955–1957

Harold
Macmillan
1957–1963

Alec Douglas-
Home
1963–1964

Harold Wilson
1964–1970

Edward Heath
1970–1974

Harold Wilson
1974–1976

James
Callaghan
1976–1979

Margaret
Thatcher
1979–1990

John Major
1990–1997

Tony Blair
1997–2007

Gordon Brown
2007–2010

David
Cameron
2010–2016

Theresa May
2016–

LEARN ABOUT THE MAJOR WARS SINCE 1899

The Second Anglo-Boer War
1899–1902

The First World War
1914–1918

The Russian Civil War
1917–1922

The Irish War of Independence
1919–1921

The Irish Civil War
1922–1923

The Second World War
1939–1945

The Malayan Emergency
1948–1960

The Korean War
1950–1953

The Kenya Emergency
1952–1960

The Suez Crisis
1956

The Aden Emergency
1963–1967

The Troubles
1968–1998

The Falklands War
1982

The Gulf War
1990–1991

The Bosnian War
1992–1995

The Kosovo War
1998–1999

The War in Afghanistan
2001–2014

The Iraq War and Insurgency
2003–2011

The Syrian Civil War
2011–

REMEMBER
IMPORTANT DATES

There are several handy rhymes to remember important dates in history – try these ones for size:

- In fourteen hundred and ninety-two Columbus sailed the ocean blue (Christopher Columbus discovers the New World)

- In fifteen hundred and eighty-eight the Spanish Armada met its fate (the Spanish Armada)

- In sixteen hundred and sixty-five scarce a soul was left alive (the Plague)

- In sixteen hundred and sixty-six London burned like dried-up sticks (the Great Fire of London)

DISCOVER PRE-1066 MONARCHS

Egbert
802–839

Aethelwulf
839–858

Aethelbald
858–860

Aethelberht
860–866

Aethelred I
866–871

Alfred the Great
871–899

Edward the Elder
899–925

Athelstan
925–940

Edmund the Magnificent
940–946

Eadred
946–955

Eadwig (Edwy) All-Fair
955–959

Edgar the Peaceable
959–975

Edward the Martyr
975–978

Aethelred II (Ethelred the Unready)
979–1013 and 1014–1016

Svein Forkbeard
1014

Edmund II (Ironside)
1016

Cnut (Canute)
1016–1035

Harold I
1035–1040

Hardicnut
1040–1042

Edward (the Confessor)
1042–1066

Harold II
1066

LEARN ABOUT TOP INVENTORS

- **Thomas Edison** – Invented the phonograph (1877) and motion picture camera (1892)

- **Wright brothers** – Built and flew the first powered aircraft (1903)

- **Benjamin Franklin** – Invented the lightning rod (1749)

- **Nikola Tesla** – Developed modern electricity as we know it (1880s–90s)

- **Konrad Zuse** – Invented the first programmable computer (1941)

- **Alexander Graham Bell** – Credited with inventing the telephone (1876)

- **Leonardo da Vinci** – Inventor of many things that proved workable 500 years later (b. 1452; d. 1519)

- **Galileo** – Developed the telescope (1609)

- **Tim Berners-Lee** – Inventor of the world wide web (1989)

LEARN THE ROMAN NAMES FOR BRITISH CITIES AND TOWNS

Aquae Sulis
Bath

Caesaromagus
Chelmsford

Camulodunum
Colchester

Corinium Dobunnorum
Cirencester

Danum
Doncaster

Deva Victrix
Chester

Dubris
Dover

Durnovaria
Dorchester

Durocornovium
Swindon

Durolipons
Cambridge

Durovernum Cantiacorum
Canterbury

Eboracum
York

Evidensca
East Lothian

Glevum Colonia
Gloucester

Isca Dumnoniorum
Exeter

Leodis
Leeds

Londinium
London

Luguvalium
Carlisle

Mamucium
Manchester

Moridunum
Carmarthen

Noviomagus Reginorum
Chichester

Pons Aelius
Newcastle

Ratae Corieltauvorum
Leicester

Venta Belgarum
Winchester

Verulamium
St Albans

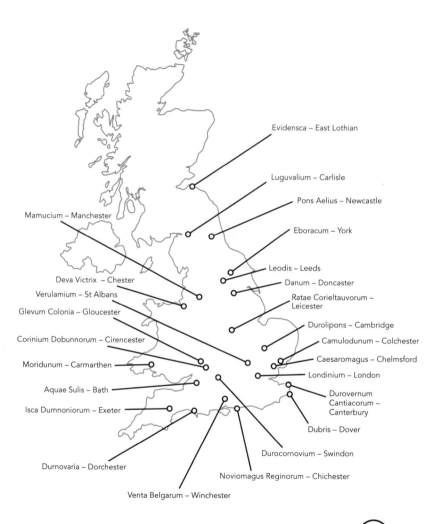

Evidensca – East Lothian

Luguvalium – Carlisle

Pons Aelius – Newcastle

Mamucium – Manchester

Eboracum – York

Leodis – Leeds

Deva Victrix – Chester

Danum – Doncaster

Verulamium – St Albans

Ratae Corieltauvorum – Leicester

Glevum Colonia – Gloucester

Corinium Dobunnorum – Cirencester

Durolipons – Cambridge

Camulodunum – Colchester

Moridunum – Carmarthen

Caesaromagus – Chelmsford

Aquae Sulis – Bath

Londinium – London

Isca Dumnoniorum – Exeter

Durovernum Cantiacorum – Canterbury

Dubris – Dover

Durnovaria – Dorchester

Durocornovium – Swindon

Noviomagus Reginorum – Chichester

Venta Belgarum – Winchester

GEOGRAPHY

Ever felt compelled to learn the shipping forecast
areas? Or all the countries in Africa? What about
all 50 states of America or the counties of England?
Or how to measure the Earth? Set yourself a few
challenges and you'll be surprised at what
you can achieve.

KNOW THE TOP 10 LARGEST COUNTRIES BY AREA

COUNTRY	AREA IN SQ KM
Russia	17,098,242
Canada	9,984,670
United States	9,826,675
China	9,596,960
Brazil	8,514,877
Australia	7,741,220
India	3,287,263
Argentina	2,780,400
Kazakhstan	2,724,900
Algeria	2,381,741

LEARN THE SHIPPING FORECAST AREAS

There's something quite comforting about the soothing sounds of the shipping forecast on the radio, even if you don't know what on earth they're talking about. Learn the shipping areas for yourself so you will next time.

Bailey	Forth	Shannon
Biscay	German Bight	Sole
Cromarty	Hebrides	Southeast Iceland
Dogger	Humber	
Dover	Irish Sea	South Utsire
Faeroes	Lundy	Thames
Fair Isle	Malin	Trafalgar
Fastnet	North Utsire	Tyne
Fisher	Plymouth	Viking
FitzRoy	Portland	Wight
Forties	Rockall	

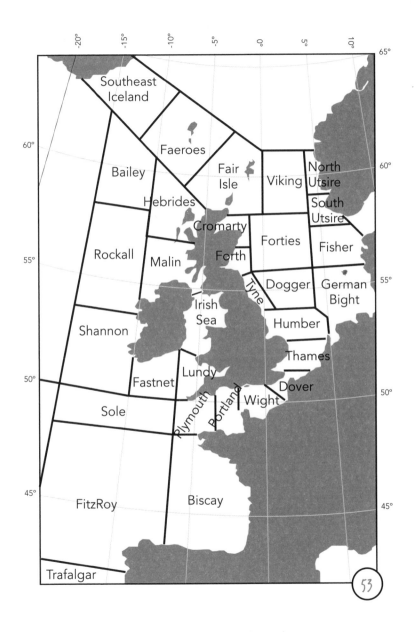

LEARN ALL 50 US STATES

ALABAMA

Alaska

Arizona

ARKANSAS

California

Colorado

CONNECTICUT

Delaware

Florida

Georgia

Hawaii

Idaho

ILLINOIS

Indiana

Kansas

KENTUCKY

IOWA

Maine

Maryland

Louisiana

Minnesota

Massachusetts

MICHIGAN

Montana

Mississippi

Missouri

NEW HAMPSHIRE

NEBRASKA

Nevada

NEW JERSEY

New Mexico

New York

Oklahoma

North Carolina

NORTH DAKOTA

Ohio

OREGON

Pennsylvania

Rhode Island

South Carolina

SOUTH DAKOTA

Tennessee

Texas

UTAH

Vermont

VIRGINIA

WEST VIRGINIA

Wisconsin

Washington

WYOMING

WORK YOUR HEAD AROUND POPULATION DENSITY

Did you know that if the entire world were as densely populated as New York City, we would collectively cover 648,543 square km (250,404 square miles)? This means that everyone would fit into the US state of Texas. On the other hand, if everyone were to be spaced out as per the population density in Houston, Texas, we would need 4.58 million square km (1.77 million square miles) in which to live.

NAME ALL OF THE CHANNEL ISLANDS

There are seven inhabited Channel Islands – Alderney, Brecqhou, Guernsey, Herm, Jersey, Jethou and Sark.

KNOW THE COUNTIES OF ENGLAND

Bath and North East Somerset

BEDFORDSHIRE

Berkshire

Bristol

BUCKINGHAMSHIRE

Cambridgeshire

County Durham

CORNWALL

Cumbria

EAST RIDING OF YORKSHIRE

Derbyshire

Devon

Dorset

East Sussex

Essex

Gloucestershire

GREATER MANCHESTER

HAMPSHIRE

Herefordshire

Hertfordshire

Isle of Wight

Kent

Lancashire

Merseyside

Leicestershire

LINCOLNSHIRE

NORTHAMPTONSHIRE

Norfolk

NORTHUMBERLAND

North Somerset

North Yorkshire

Shropshire

Nottinghamshire

OXFORDSHIRE

Rutland

SOMERSET

South Gloucestershire

Suffolk

South Yorkshire

STAFFORDSHIRE

TYNE AND WEAR

Surrey

WEST MIDLANDS

Wiltshire

West Sussex

WARWICKSHIRE

West Yorkshire

WORCESTERSHIRE

LEARN ABOUT ANTARCTICA

The ice sheet in Antarctica spans a whopping 14 million square km (5.4 million square miles) and is the largest solid ice mass on the planet, also making it the home of almost all of the world's fresh water.

MEASURE THE EARTH

The circumference of the Earth around the equator is 40,070 km (24,898 miles). However, if you were to measure the planet's circumference the other way, crossing over the poles, the measurement would be a little less, at 39,931 km (24,812 miles). This is due to the flattening of the Earth at the poles, creating a shape called an oblate spheroid.

KNOW YOUR CONTINENTS

CONTINENT	NO. OF COUNTRIES	FACT
Asia	50	60 per cent of Earth's population live here
Africa	55	This is the hottest continent, with the largest desert (the Sahara), which covers 25 per cent of Africa.
North America	32	The USA is the largest economy in the world
South America	12	Contains the largest forest (the Amazon rainforest), which covers 30 per cent of the continent
Antarctica	1	The coldest continent in the world
Europe	51	The EU is the biggest economic and political union in the world
Australasia or Oceania	14	The least populated continent after Antarctica (only 0.3 per cent of the world's population lives here)

LEARN THE COUNTRIES
OF AFRICA

Algeria

Angola   Benin   BOTSWANA   Burkina Faso

Burundi   CAMEROON   Cape Verde

Central African Republic   Chad   Comoros

Djibouti   Eritrea

DEMOCRATIC   Equatorial Guinea   Ethiopia
REPUBLIC OF THE
CONGO   GAMBIA   Ghana

Gabon   Côte d'Ivoire   LIBERIA

Guinea

Guinea Bissau   Lesotho

Madagascar   Mauritania

LIBYA
Mali   MOROCCO   Namibia
MALAWI   Mozambique

Mauritius   Nigeria

RWANDA   Republic of the Congo

NIGER   SENEGAL

Réunion   SIERRA LEONE   Somalia

Seychelles   Sao Tome & Principe

SOUTH AFRICA   Sudan   SWAZILAND   Tanzania

Togo   Tunisia   UGANDA   Western Sahara

Zambia   ZANZIBAR   Zimbabwe

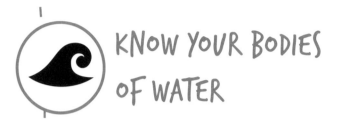

KNOW YOUR BODIES OF WATER

There are five oceans on Earth – the Atlantic, Arctic, Indian, Pacific and Southern oceans – but the term 'seven seas' dates back to ancient times and comes from the splitting of the Atlantic and Pacific into the North Atlantic and South Atlantic and North Pacific and South Pacific. When it comes to the other bodies of water across the globe, they include seas, rivers, gulfs, bays, lakes, canals, falls and straits, and there are too many to count!

INTO THE DEEP

The deepest point on Earth is in the western Pacific, at the southern end of the Mariana Trench, where the ocean reaches depths of 11,034 metres (36,200 feet).

WORDS AND LANGUAGE

Rediscover the dying art of postcard writing, learn how to have a bit of banter in Cockney rhyming slang and pick up Pig Latin to have secretive chats with friends. Better still, make up your own language.

TRY COCKNEY RHYMING SLANG

Adam and Eve
Believe

Apples and pears
Stairs

April showers
Flowers

Barnet Fair
Hair

Battle cruiser
Boozer

Boat race
Face

Brown bread
Dead

Butcher's hook
Look

China plate
Mate

Cream crackered
Knackered

Daisy roots
Boots

Giraffe
Laugh

Jack and Jill
Hill

Lady Godiva
Fiver

Plates of meat
Feet

Pork pies
Lies

Rosie Lee
Tea

Ruby Murray
Curry

Scooby Doo
Clue

Sherbet dab
Cab

Titfer (tit for tat)
Hat

DEGREE IN COCKNEY

In Cockney rhyming slang university degrees are referred to as Geoff Hurst (first), Desmond Tutu (2:2) and Douglas Hurd (third). If you get a 2:1 you have done brilliantly, but sadly have no CRS term for your achievement.

WRITE FLASH FICTION

This is an excellent exercise to get your creative juices flowing and can be done with friends or as a solo mission (although it can be more fun to have people to share the end product with). Set yourself a subject or first line of the story and a time or word limit – an hour or 500 words, whichever you reach first, is a good place to start. Take yourself away from your daily tasks and flex your storytelling muscles.

COMPOSE A HAIKU

A haiku is a Japanese poem that uses sensory language to describe a feeling or experience. They are often inspired by nature, do not rhyme, always consist of three lines and the syllables always follow the same pattern. The first line should consist of five syllables, the second line seven syllables and the third line five syllables again. Here's an example:

An old silent pond...
A frog jumps into the pond,
splash! Silence again.

Bashō

SEND A POSTCARD

Whether you're on holiday, on a day trip or just pottering about locally, send a postcard to someone to show them you're thinking about them. The traditional postcard will always be the best option, but if you want to add a more personal touch you could get postcards printed online using your own photos.

 # LEARN ROMAN NUMERALS

1 I	9 IX	17 XVII	300 CCC
2 II	10 X	18 XVIII	400 CD
3 III	11 XI	19 XIX	500 D
4 IV	12 XII	20 XX	600 DC
5 V	13 XIII	40 XL	700 DCC
6 VI	14 XIV	50 L	800 DCCC
7 VII	15 XV	100 C	900 CM
8 VIII	16 XVI	200 CC	1,000 M

SPEAK PIG LATIN

Pig Latin is used by English-speaking people the world over in situations where they might want to hide something from eavesdroppers – but it will only work if the eavesdroppers don't speak Pig Latin themselves! Simply move the first letter of the word to the end and add 'ay'. If the word begins with a vowel leave it as it is and add 'way'. Here are some examples:

Hello – ellohay

Banana – ananabay

Pig Latin – igpay atinlay

Batman and Robin – atmanbay andway obinray

# LEARN MORSE CODE

A	.−	S	...
B	−...	T	−
C	−.−.	U	..−
D	−..	V	...−
E	.	W	.−−
F	..−.	X	−..−
G	−−.	Y	−.−−
H	....	Z	−−..
I	..	1	.−−−−
J	.−−−	2	..−−−
K	−.−	3	...−−
L	.−..	4	....−
M	−−	5	.....
N	−.	6	−....
O	−−−	7	−−...
P	.−−.	8	−−−..
Q	−−.−	9	−−−−.
R	.−.	0	−−−−−

Here is some morse code for you to try your hand at:

a. -.-. --- -. --. .-. .- - ..- .-.. .- - .. --- -. ... /
-.-- --- ..- / -.-. .-. .- -.-. -.- . -.. / - .... . /
-.-. --- -.. .

b. -- --- .-. ... . / -.-. --- -.. . / .-- .- ... / .. -.
...- . -. - . -.. / .. -. / .---- ---.. ...-- -.... /
-... -.-- / .... . -. .-. -.-- / -- --- .-. ... .

c. -- --- .-. ... . / -.-. --- -.. . / -.-. .- -. / -....
/ - .-. .- -. ... -- .. - - . -.. / --- ...- . .-. / .-.
.- -.. .. --- / --- .-. / -... -.-- / ..-. .-.. .- ... .
.. -. --. / .-.. .. --. .... - ...

FAMOUS MESSAGES

The most famous messages to be sent via Morse code are 'V' (. . . -) for victory, which was sent at the end of World War Two, and 'SOS' (. . . - - - . . .).

KNOW YOUR LONDON TAXI DRIVER SLANG

Bowler hat
City worker

Bullseye
£50 note

Burst
Pub and club chucking out time

Cage
Passenger compartment

Cargo
Passengers

Cockle
£10 note

Confessional
Pull-down seat behind driver

The Dilly
Piccadilly

Dogfight
Battle with traffic

Drop heavy
Passenger who tips well

Flyer
Job to Heathrow

Gas works
Houses of Parliament

Grease up
Stop for food

Honeypot
West End

Magic carpet
Pedestrian crossing

Mooch
Look for work

The Saveloy
The Savoy Hotel

Score
£20 note

Shock absorbers
Multiple passengers

The Uproar
Royal Opera House

CREATE YOUR OWN LANGUAGE

A good way to create your own language and then communicate in code among your friends or family is to mix up the alphabet. Come up with a pattern you could use to assign each letter of the alphabet to a new counterpart, for example by flipping the alphabet, so A = Z, B = Y, C = X, etc. Alternatively, you could jump ahead two letters, so A = C, B = D, C = E, etc.

KNOW YOUR AMERICAN ENGLISH

US English
UK English

Barrette
Hair slide

Beet
Beetroot

Cilantro
Coriander

Chips
Crisps

College
University

Cotton candy
Candyfloss

Drugstore
Chemist

Eggplant
Aubergine

Elementary school
Primary school

Elevator
Lift

Emergency room
A&E

First floor
Ground floor

Fish stick
Fish finger

French fries
Chips

Gas
Petrol

Highway
Motorway

Ladybug
Ladybird

Monkey wrench
Spanner

Purse
Handbag

Pants
Trousers

Pacifier
Dummy

Popsicle
Ice lolly

Scallion
Spring onion

Second floor
First floor

Station wagon
Estate car

Thumbtack
Drawing pin

Trunk
Car boot

Vacation
Holiday

Vest
Waistcoat

Zip code
Postcode

Zucchini
Courgette

LEARN SEMAPHORE

 A

 B

 C

 G

 H

 I

 M

 N

 O

 S

 T

 U

 **Y**

 Z

KEEP THE ART OF LETTER WRITING ALIVE

The art of letter writing is fast being lost, with emails, text messages and social media platforms quickly taking over as far swifter forms of communication. Make a point of sending a letter every once in a while and help to keep the practice alive. The recipient will delight in receiving a handwritten or typed missive from you and will most likely sit down and take time out from their day to read about your news.

FOOD

Whether you've always wanted to know how to grow your own name, master chopsticks or chop onions without crying, it's all here. While you're at it, learn all the plant food families and different pastas, and the very best way to crack a coconut.

PICKLE ONIONS

Take 500 g of small shallots, 50 g of salt, 500 ml of malt vinegar and 200 g of clear honey. Place the shallots in a bowl, cover with boiling water and leave to cool. Fish them out with a slotted spoon and trim the roots and tops, then return to the water and sprinkle in the salt. Leave overnight. Rinse the onions and pat dry. Mix the honey and vinegar in a large pan and heat but don't boil. Place the onions into sterilised jars and pour over the honey and vinegar mixture. Add dried chillies/peppercorns for a kick, seal the jars and leave to cool. The onions will be ready to eat in one month.

KNOW YOUR PLANT FOOD FAMILIES

Umbelliferae – carrots, celery, fennel, parsnips, parsley

Lamiaceae – basil, mint, oregano, sage, thyme

Solanaceae – aubergines, peppers, potatoes, tomatoes

Asteraceae – artichokes, chicory, tarragon, lettuce

Brassicaceae – broccoli, cabbages, cauliflower, watercress

Liliaceae – asparagus, garlic, leeks, onions

Rosaceae – apples, blackberries, raspberries, strawberries

Cucurbitaceae – cucumbers, melons, pumpkins, squashes

Chenopodiaceae – beetroots, chard, spinach

Fabaceae – beans, peas, lentils

Poaceae – barley, corn, oats, rice, wheat

GROW YOUR NAME

On a flat piece of damp cotton or gauze, sprinkle small seeds (cress seeds are ideal) in the shape of the letters of your name. Keep the cotton moist, and in a few days your name will start to grow. An ideal gift for someone who might need cheering up or something for your own kitchen windowsill!

MASTER CHOPSTICKS

Rest the thicker end of one of the chopsticks in the fleshy part of your hand between the base of your thumb and forefinger and hold the lower part of the chopstick between your middle and ring fingers. Place the other chopstick on top and hold between your thumb and forefinger, gripping it halfway along. You should now be able to make a pincer motion with the two chopsticks that will enable you to pick up bits of food and feed yourself. As they say in China, *chī hǎo hē hǎo* (enjoy your meal)!

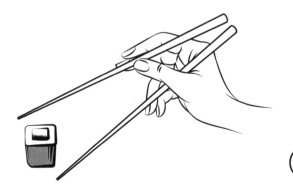

MAKE CHOCOLATE FRIDGE CAKE

1. Melt 300 g of dark chocolate in a glass bowl over a saucepan of boiling water.

2. Add 100 g of butter and 150 g golden syrup and stir until smooth.

3. In a separate bowl crush 250 g of digestive biscuits.

4. Add a handful of raisins, glacé cherries and nuts or mini marshmallows to the biscuits if you desire.

5. Mix the wet and dry mixtures together and pack into a lined square tin.

6. Place in the fridge for a few hours then remove and cut into squares.

MAKE COOKIES LAST LONGER

Put half a slice of bread into the tin; the cookies will absorb the moisture from the bread and stay soft and chewy for longer (make sure to replace the bread after a day or so).

ROAST SWEET CHESTNUTS

There are a few different ways to roast chestnuts and the results are all equally pleasing. Whichever method you choose, be sure to prepare the chestnuts before you cook them – either carefully pierce the skin with a skewer or fork, or cut a slit across the dome of the nut. Then, either shake the nuts in a pan on the hob or place in a dish with a little water and roast for 10 minutes. Add salt or seasoning of your choice and enjoy the sumptuous taste of winter.

CRACK A COCONUT

Ever won a coconut at the summer fete and then found yourself clueless as to how to get at the juicy white flesh? Aside from the chuck-it-at-a-wall method, which isn't the safest or the cleanest, here's how to savour all that lovely coconut without losing an inch of it. Pierce the eyes of the coconut and drain out the milk, then place it in the oven at 170°C for 30 minutes. Leave to cool. If the coconut hasn't cracked on its own, give it a light tap with a hammer.

KEEP FLIES OUT OF YOUR KITCHEN

Grow pots of basil or leave bay leaves on your windowsill, and either grow or leave bunches of lavender about the place, and flies will stay away.

DRY CHILLIES

If you have a bumper crop of chillies, a good way to preserve them for use all year is to dry and string them. They can also look rather colourful in your kitchen. Select as many chillies as you like, and wash and dry them. Thread a needle with flavour-free dental floss or strong thread and make a loop at one end. Insert the needle into the chilli near the stem and push it all the way to the loop, then tie a knot. Add as many chillies as you like and hang up in the window. Wait at least three weeks before you use them. Either grind them up and use as chilli powder or chop up roughly and add to a bottle of olive oil.

HOW TO MAKE YOUR DRINK COLD QUICKLY

Take a piece of paper towel, run it under
the tap to dampen it, wrap it around your
drink and place in the freezer for
15 minutes. You'll be amazed
at how quickly it cools down.

MAKE AN EGG STAND UP

Now for a bit of kitchen 'magic'. Take an egg and shake it vigorously for 10 seconds to mix up the white and the yolk. Then place the egg on a surface so it is standing upright on its wider end and hold it there for several minutes. During this time the heavier yolk sinks to the bottom of the egg to offer a sturdy base and you will find that eventually the egg stands upright on its own.

KNOW YOUR PASTAS AND HOW TO SERVE THEM

Campanelle – Serve with thick, creamy sauces

 Cannelloni – Stuff with meat or veggies and cover in white and tomato sauce

Conchiglie – Stuff the big ones, add the medium-sized to casseroles and the smallest to soups

 Farfalle – Make pasta salad or serve with meat and vegetables

Fusilli – Works with all manner of toppings and sauces

 Gnocchi – Serve with rich sauces

Linguine – Use as a spaghetti alternative, serve with seafood or chunky tomato sauce

Macaroni – Performs best in the much-loved macaroni cheese; add pesto for a greener version

Orzo – Add to soups and salads, or serve hot with vegetables

Pappardelle – Serve with hearty, often meaty, sauces

Penne – Very versatile and goes with most sauces and toppings

Ravioli – Stuff with almost anything and keep it simple with a drizzle of olive oil, or top with cheese or tomato sauce

MAKE YOUR OWN GRANOLA

Preheat the oven to 150°C and line a baking tray with greaseproof paper. In a large bowl mix 390 g oats, 75 g sunflower seeds, 75 g almond slices/flakes, 75 g dried fruit, ½ tsp cinnamon, ½ tsp salt, 80 ml olive oil and 120 ml maple syrup. Lay the mixture on the baking tray and bake for 40 minutes, turning it over every 10 minutes. Leave to cool and store in a jar.

CHOP ONIONS WITHOUT CRYING

There are several methods you can try to avoid those onion-induced tears. Try freezing the onion for half an hour or submerging it in cold water for a while before you slice it. If that doesn't work, there are specially designed goggles for the job. Slightly wackier but tried and tested: chop the onion with a piece of bread hanging out of your mouth – the bread will absorb the onion gases that cause the tears to flow before they have a chance to reach your eyes.

SPORT AND FITNESS

Learn how to hold a yoga pose and bring about some calm by practising meditation. There's plenty of trivia here crossing the sports spectrum, along with tips on how to build up to a 5k run and learn some new swimming strokes.

HOLD TREE POSE

This one requires a bit of balance, but you might say it's the 'classic' yoga pose.

1. Stand with your feet together and your hands in prayer position.

2. Spread your toes and press them firmly into your mat/the ground.

3. Take one of your feet off the ground and rest the sole either against your other ankle or your inner thigh, if you can manage it, toes pointing to the floor.

4. Keep your breath steady and try not to wobble. Staring at a fixed point can help you to keep your balance.

TAKE YOUR OWN PULSE (OR SOMEONE ELSE'S)

Hold out one of your arms, palm up, elbow slightly bent. Using your other hand, place your index and middle finger on the inside of your wrist, at the base of your thumb. Press lightly until you can feel your pulse. Count the number of beats for one minute (or for 30 seconds and multiply it by two). To find your pulse in your neck, press the same fingers on to the side of your neck in the soft hollow below your jaw, next to your windpipe. A healthy resting heart rate for non-athletes should sit between 60 and 100 beats per minute.

LEARN THE TOP TEN MOST DECORATED OLYMPIANS

RANK	NAME	COUNTRY	SPORT
1	Michael Phelps	USA	Swimming
2	Larisa Latynina	Soviet Union	Gymnastics
3	Nikolai Andrianov	Soviet Union	Gymnastics
4	Ole Einar Bjørndalen	Norway	Biathalon
5	Boris Shakhlin	Soviet Union	Gymnastics
6	Edoardo Mangiarotti	Italy	Fencing
7	Takashi Ono	Japan	Gymnastics
8	Paavo Nurmi	Finland	Athletics
=9	Birgit Fischer	East Germany/ Germany	Canoeing
=9	Bjørn Dæhlie	Norway	Cross-country skiing

BUILD UP TO A 5K RUN

Training should take place every other day, three times a week.

Week 1

Brisk 5-minute walk; 1 minute running alternated with 1.5 minutes walking for a total of 20 minutes

Week 2

Brisk 5-minute walk; 1.5 minutes running alternated with 2 minutes walking for a total of 20 minutes

Week 3

Brisk 5-minute walk; two rounds of 1.5 minutes running, 1.5 minutes walking, 3 minutes running, 3 minutes walking

Week 4

Brisk 5-minute walk, 3 minutes running, 1.5 minutes walking, 5 minutes running, 2.5 minutes walking, 3 minutes running, 1.5 minutes walking and 5 minutes running

Week 5

Run one: Brisk 5-minute walk, 5 minutes running, 3 minutes walking, 5 minutes running, 3 minutes walking and 5 minutes running

Run two: Brisk 5-minute walk, 8 minutes running, 5 minutes walking and 8 minutes running

Run three: Brisk 5-minute walk, 20 minutes running with no walking

Week 6

Run one: Brisk 5-minute walk, 5 minutes running, 3 minutes walking, 8 minutes running, 3 minutes walking and 5 minutes running

Run two: Brisk 5-minute walk, 10 minutes running, 3 minutes walking and 10 minutes running

Run three: Brisk 5-minute walk, 25 minutes running with no walking

Week 7

Brisk 5-minute walk, 25 minutes running

Week 8

Brisk 5-minute walk, 28 minutes running

Week 9

Brisk 5-minute walk, 30 minutes running

TOP TEN GOALSCORERS IN THE PREMIER LEAGUE

RANK	NAME	GOALS
1	Alan Shearer	260
2	Wayne Rooney	202
3	Andrew Cole	187
4	Frank Lampard	177
5	Thierry Henry	175
6	Robbie Fowler	163
7	Jermain Defoe	159
8	Michael Owen	150
9	Les Ferdinand	149
10	Teddy Sheringham	146

EAT A BREAKFAST OF CHAMPIONS

At the height of his career, the most decorated Olympian – swimmer Michael Phelps – was consuming enormous breakfasts to get him through his intensive training regime. Imagine chowing down on the following every day!

- Three fried egg sandwiches with cheese, lettuce, tomatoes, fried onions and mayonnaise
- Two cups of coffee
- One five-egg omelette
- One bowl of cereal
- Three slices of French toast
- Three chocolate-chip pancakes

Phelps went on to tweak this to a vat of porridge and a large omelette, followed by fruit and coffee.

MEDITATE FOR TEN MINUTES

Meditation is a practice that can help us to relax and focus on the present, driving away unwanted thoughts and worries.

1. Find a quiet space where you won't be interrupted.

2. Sit up straight on the floor with your legs crossed, or in a chair, with your hands on your knees.

3. Close your eyes.

4. Breathe deeply, in through your nose and out through your mouth.

5. Concentrate only on your breath and notice the sensations.

6. If you begin to think about something else, gently steer yourself back to your breath.

7. Do this for 10 minutes every day or whenever you can to feel calm and collected.

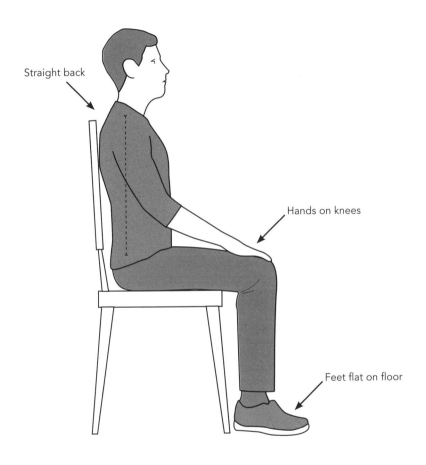

Straight back

Hands on knees

Feet flat on floor

LEARN SOME NEW SWIMMING STROKES

Most of us will be familiar with front crawl, breaststroke and backstroke, and some of us will even be pretty nifty at butterfly, but why not try one of these styles of swimming to mix things up?

Combat sidestroke – Used by US Navy Seals to swim efficiently and minimise their profile in the water, this involves stretching both arms out in front of you while flutter kicking your legs (as you would in front crawl); your left arm should pull through the water beside you, followed by your right arm, which will automatically tilt you to the side to breathe; at the same time scissor-kick your legs as you would in breaststroke; then reach both arms in front of you again to begin the next stroke.

Trudgen stroke – Used by swimmers to develop both upper- and lower-body strength, this involves stretching each arm forward alternately in front of you, a bit like front crawl but keeping your head out of the water, and scissor-kicking your legs (like you would in breaststroke).

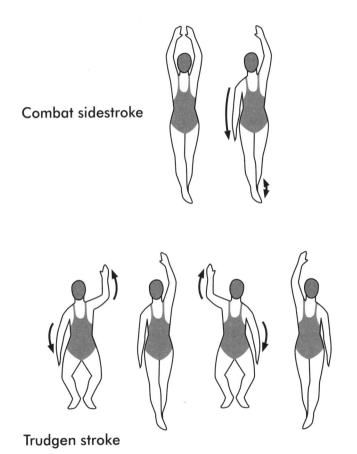

Combat sidestroke

Trudgen stroke

LEARN THE TOP 100 M SPRINTERS

Men

RANK	NAME	NATIONALITY	TIME
1	Usain Bolt	Jamaica	9.58 seconds
2	Tyson Gay	United States	9.69 seconds
3	Asafa Powell	Jamaica	9.72 seconds
4	Maurice Greene	United States	9.79 seconds
=5	Donovan Bailey	Canada	9.84 seconds
=5	Bruny Surin	Canada	9.84 seconds

Women

RANK	NAME	NATIONALITY	TIME
1	Florence Griffith-Joyner	United States	10.49 seconds
2	Carmelita Jeter	United States	10.64 seconds
3	Marion Jones	United States	10.65 seconds
4	Shelly-Ann Fraser	Jamaica	10.73 seconds
5	Christine Arron	France	10.73 seconds

LEARN THE TOP TENNIS GRAND SLAM WINNERS

NAME	COUNTRY	TOTAL WINS
Margaret Court	Australia	24
Serena Williams	USA	23
Steffi Graf	Germany	22
Roger Federer	Switzerland	19
Helen Wills Moody	USA	19
Chris Evert	USA	18
Martina Navratilova	USA	18
Rafael Nadal	Spain	15
Pete Sampras	USA	14
Novak Djokovic	Serbia	12
Roy Emerson	Australia	12
Billie Jean King	USA	12

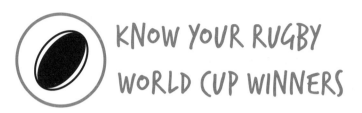

KNOW YOUR RUGBY WORLD CUP WINNERS

YEAR	WINNER	SCORE	RUNNER-UP
1987	New Zealand	29–9	France
1991	Australia	12–6	England
1995	South Africa	15–12	New Zealand
1999	Australia	35–12	France
2003	England	20–17	Australia
2007	South Africa	15–6	England
2011	New Zealand	8–7	France
2015	New Zealand	34–17	Australia

STEP IT UP

Wearable fitness trackers typically recommend a daily goal of 10,000 steps to keep fit and healthy, and this roughly translates to around 5 miles. If that's too much for you, set your sights a bit lower and try to achieve your goal each day before working your way up. The benefits of walking are many, with fresh air and exercise not only being essential to general health and well-being, but also likely to have a positive effect on your mood.

THE ARTS

Whether you've always wanted to master the nose flute or learn how to make one of those cute origami birds, look no further. Master the art of face drawing, make marbled paper and memorise the best-selling albums of all time while clicking away on the castanets.

LEARN THE FIVE MOST EXPENSIVE WORKS OF ART SOLD AT AUCTION

ARTIST	WORK OF ART	SALE PRICE	AUCTION DATE
Leonardo da Vinci	*Salvator Mundi (Saviour of the World)*	$450 million	2017
Pablo Picasso	*Les Femmes d'Alger (Women of Algiers)*	$179 million	2015
Amedeo Modigliani	*Nu Couché (Reclining Nude)*	$170 million	2015
Francis Bacon	*Three Studies of Lucian Freud*	$142 million	2013
Edvard Munch	*Skrik (The Scream)*	$120 million	2012

MAKE AN ORIGAMI BIRD

1. Take a square piece of paper (origami paper will look the nicest) and fold the top corner to the bottom corner.

2. Now fold the left corner over to the right corner.

3. Open the top flap and fold the top corner to meet the bottom corner; then flip over and do the same on the other side.

4. On one side, fold the left and right corners to meet in the centre and fold down the top triangle.

5. Unfold step 4 and open up the top layer, using the creases as guides fold the left and right sides into the centre. Do the same on the other side.

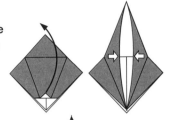

6. Take the top layer from the right side and fold it to the left. Do the same on the other side.

7. Take the bottom corner of the top layer and fold it up. Do the same on the other side.

8. Take hold of the two wedges in the middle and pull them out to the sides. Crease along the bottom to hold these in place.

9. Open the left wedge and fold it in on itself to create the bird's head, fold the wings down and voilà!

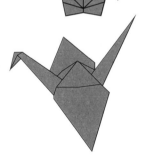

DRAW A FACE

1. Using a pencil, draw the shape of the head.

2. Halfway from the top, lightly draw a horizontal line (this is where the eyes will go).

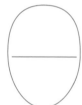

3. Lightly draw a vertical line down the middle (this will help to keep the features in the right place).

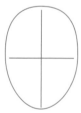

4. Halfway between the eye line and the chin, draw a short line (this will become the nose).

5. Slightly above the halfway point between the nose and chin draw a longer line (this will become the mouth).

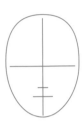

6. Using an art pencil or pen for the next steps, draw two curves on the eye line and then study the eyes you are drawing and adjust the shape accordingly below the line. See the example on this page.

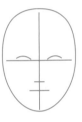

7. Draw on a nose and mouth using your guide lines. Then add hair and remember it needs to sit above and below the skull outline.

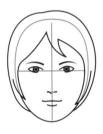

8. Erase all your pencil guidelines and there you have it.

MASTER THE NOSE FLUTE

You don't need much musical experience to play this instrument, so why not give it a try? Press the smaller end of the flute against your nose and the larger end against your mouth. Form a seal with the instrument around your nostrils, but keep your mouth open slightly. Only breathe out through your nose and as you begin to create tunes, alter the shape of your mouth as you exhale through your nose to vary the pitch.

LEARN TO PLAY
THE CASTANETS

Do your best flamenco impression with these easy-to-follow instructions. Take the castanets and place one in each hand. Thread your thumb through the loop so your thumb sits on the outside and you can use your other fingers to click the plates together. Your left hand should play a simple rhythm while your right takes on the more complicated rhythm of the dance. See how you get on!

CHANGE THE COLOUR OF A FLOWER

Ever had a hankering for a bunch of blue roses but never see them at the local florist? Well here's how you can make your own. Buy some pale roses – white or cream will work best – and place the stems into a strong solution of blue ink or food colouring and water (or whatever colour you desire). The rose will 'drink' the ink and eventually change colour. Try making a multi-coloured rose by splitting the stem in half and placing each half into a different-coloured solution.

TAKE A GREAT PHOTOGRAPH

Rediscover the lost art of taking a
photograph with these handy tips.

Focus – Make sure your subject matter is
looking sharp.

Lighting – Always think about the light.
Sunlight overhead can cast shadows and the
sun behind you will make people squint. Play
around with different positions to find the
perfect spot.

Horizon – Always keep it straight!

Perspective – Create a sense of depth
and perspective by including items in the
foreground of your photograph.

Natural lines – Whether these are in the
clouds, trees, roads or buildings, horizontal,
vertical or zigzag, lines will guide the
viewer's eyes around your photograph.

LEARN THE TOP 10 ALBUMS OF ALL TIME

ARTIST	ALBUM	YEAR	SALES
Michael Jackson	*Thriller*	1982	66 million
AC/DC	*Back in Black*	1980	50 million
Pink Floyd	*The Dark Side of the Moon*	1973	45 million
Whitney Houston/ Various artists	*The Bodyguard*	1992	45 million
Meat Loaf	*Bat Out of Hell*	1977	43 million
Eagles	*Their Greatest Hits (1971–1975)*	1976	42 million
Bee Gees/ Various artists	*Saturday Night Fever*	1977	40 million
Fleetwood Mac	*Rumours*	1977	40 million
Shania Twain	*Come On Over*	1997	39 million
Led Zeppelin	*Led Zeppelin IV*	1971	37 million

KNOW YOUR DISCS

No one can ever quite remember how many albums an artist needs to sell to be awarded a gold disc, say. Learn these and wonder no more. In the USA, Gold = 500,000 albums sold, Platinum = 1 million and Diamond = 10 million, whereas in the UK Silver = 60,000, Gold = 100,000 and Platinum = 600,000.

CREATE MARBLED PAPER

1. Take a large foil tray or casserole dish – something that will fit your desired paper size.

2. Add water to the tray to fill it 1–2 cm from the bottom.

3. Take a few different colours of nail varnish or ink and add a few drops of each to the surface of the water.

4. Using a pin or knitting needle, move the water around gently to create a marbling effect.

5. Place your paper carefully onto the surface of the water – don't let it go under.

6. Take a corner of the paper and gently lift it away from the water.

7. Leave it to dry and then admire your creation!

KNOW YOUR RECENT TURNER PRIZE WINNERS

DATE	ARTIST	MEDIUM
2008	Mark Leckey	Sculpture, film, sound, performance
2009	Richard Wright	Site-specific painting
2010	Susan Philipsz	Sound installation
2011	Martin Boyce	Installation
2012	Elizabeth Price	Video
2013	Laure Prouvost	Installation, collage, film
2014	Duncan Campbell	Video
2015	Assemble	Architecture and design
2016	Helen Marten	Installation
2017	Lubaina Himid	Installation

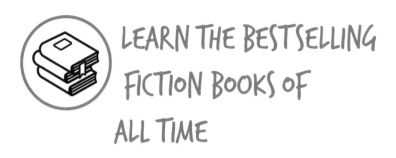

LEARN THE BESTSELLING FICTION BOOKS OF ALL TIME

TITLE	AUTHOR	SALES
Don Quixote (1605)	Miguel de Cervantes	500 million
A Tale of Two Cities (1859)	Charles Dickens	200 million
The Lord of the Rings (1954–5)	J. R. R. Tolkien	150 million
The Little Prince (1943)	Antoine de Saint-Exupéry	140 million
Harry Potter and the Philosopher's Stone (1997)	J. K. Rowling	120 million
The Hobbit (1937)	J. R. R. Tolkien	100 million
And Then There Were None (1939)	Agatha Christie	100 million
Dream of the Red Chamber (1754)	Cao Xueqin	100 million
Alice's Adventures in Wonderland (1865)	Lewis Carroll	100 million
The Lion, the Witch and the Wardrobe (1950)	C. S. Lewis	85 million

ETIQUETTE

From getting the knack of tying a bow tie, to walking in heels, dressing appropriately for an occasion and laying the table for the perfect dinner party, you can learn it all here.

WEAR FLOWERS IN YOUR HAIR

Take a plain, thin Alice band, preferably one that matches the colour of your hair, and customise it to your heart's desire. Create a flower crown with three or four large flowers in a row in the centre, or opt for a day-at-the-races-style fascinator with a stylish large white flower off to the side with a bit of black netting underneath to add a touch of class.

MAKE BUNTING FROM OLD CLOTHES

You can do this with any items of clothing, but you might find dresses or jazzy shirts offer the most suitable material as well as interesting patterns. Cut triangles out of your clothing and sew on to a long strip of bunting tape, folding the tape over the top of the short side of the triangles as you sew. For fuller bunting, sew the triangles together inside out and then reverse before sewing to the bunting tape. String up for an instant party vibe.

TIE A BOW TIE

The bow tie originated in Croatia in the seventeenth century to hold one's shirt collar together.

1. Place the bow tie round your neck. The left side should be about an inch longer than the right. Cross the left side over the right side.

2. Wrap the longer side of the bow tie around the shorter, and up through the loop. Hold it up straight.

3. Fold the shorter end of the tie in half horizontally.

4. Bring the longer end of the tie down in front of the shorter, folded end.

5. Insert it through the loop at the back of the shorter section of tie.

6. Pull both ends gently to tighten and adjust the knot.

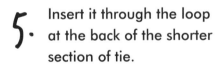

LAY A TABLE FOR A DINNER PARTY

There is a lot to remember when it comes to laying the table for a posh dinner party.

1. First start with the plate. It's likely you will plate up in the kitchen, but it's best to have something to work around.

2. Lay forks to the left of the plate and knives to the right in the order they will be used, working inwards. Place the salad knife and fork first.

3. Next place the meat fork on the left and the meat knife on the right, followed by the fish fork and knife.

4. Place the dessert fork above the plate in a horizontal position with the handle facing left and the dessert spoon above the fork with the handle facing right.

5. Finally add a side plate for bread to the left of the forks and water and wine glasses in the right-hand corner of the place setting.

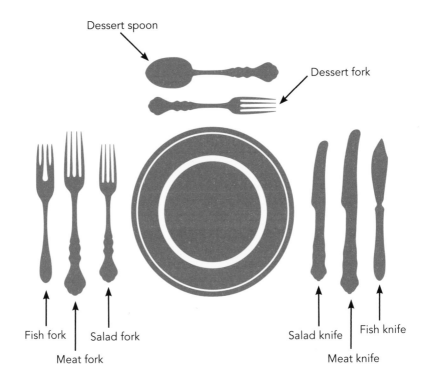

Dessert spoon

Dessert fork

Fish fork

Meat fork

Salad fork

Salad knife

Meat knife

Fish knife

 WALK IN HEELS

Straighten your back and place your heel on the ground first, followed by your toes. When your weight is on the ball of your foot, shift your weight forwards as if you are walking on your tiptoes, push forwards and then take your next step.

LEARN ABOUT TIPPING PRACTICES

Tipping practices are different across the world. In the UK, a service charge (which is not counted as a tip) is often not automatically added, and the traditional 'tip' is 10 per cent, to be shared between the party. Some restaurants specify on the receipt if an obligatory service charge is included in the bill; in these cases it is usually 12.5 per cent, and your tip is a bonus for the staff on top of that. When you are abroad, make sure you are familiar with the dos and don'ts of the country's service culture as these can often differ wildly from your own. You don't want to be chased down the street for tipping incorrectly!

 # WEAR A SILK SCARF

1. Fold in half to create a triangle, roll over the top and tie around your neck at the back, cowboy/girl style.

2. Roll it out and tie around your neck with the knot at the side, for Parisian chic.

3. Roll it up and place on your head like an Alice band, tying it underneath your loose hair, for a retro 1950s look.

4. Fold as in no. 1 and place over your head, tying under your hair for a 1990s vibe.

5. Reverse number 1 so the scarf is tied at the front, for a country living feel.

DRESS FOR THE OCCASION

Black tie – A black dinner jacket, matching trousers, a white shirt and black bow tie; or an evening gown, traditionally floor-length.

Business formal – A suit or smart shirt and formal trousers or skirt, with discreet, if any, accessories.

Smart/casual – Generally means to dress in casual clothes, but with none of the following: T-shirts, denim, trainers or beachwear.

SEND THANK-YOU NOTES

Within a couple of days of attending a wedding, party, or any kind of event hosted by someone else, make sure you send a thank-you note. These days many people will opt for a thank-you text message or note via social media, but let's keep the art of letter writing alive by popping something in the post. Have a box of thank-you notes or cards ready for such occasions (or make your own!) and the act will always be remembered.

 # MAKE MOCKTAILS

Virgin Mojito

Ingredients

5 mint leaves plus extra to garnish
1 tbsp fresh lime juice
1 tsp sugar
Crushed ice
250 ml ginger ale, soda water or lemonade
Slice of lime to garnish

Method

Place the mint leaves, lime juice and sugar in a tall glass and muddle well. Add a spoonful or two of crushed ice and top up with ginger ale, soda water or lemonade, depending on how sweet you want the end result to be. Give the whole lot a good stir and then top up with more crushed ice if needed and a sprig of mint and a slice of lime to garnish.

Mango Fizz

Ingredients

5 mint leaves plus extra to garnish
25 ml mango purée
25 ml lime juice
25 ml sugar syrup
Crushed ice
125 ml ginger ale
Slice of lime to garnish

Method

Place the mint, sugar syrup and mango purée in a
tall glass and muddle gently. Add a handful or two of
crushed ice, until the glass is two-thirds full, and top
up with ginger ale. Give the whole thing a good stir,
add more crushed ice if needed and garnish with a
sprig of mint and a slice of lime.

BE THE PERFECT HOST

- On arrival, take coats and offer guests a drink

- Offer non-drinkers their libation in a wine glass where appropriate

- Make sure there are snacks or canapés to hand to go with the first drink

- Ensure you have prepared sufficiently that you do not spend all your time in the kitchen and that the time between courses does not drag

- Keep music at a light volume so voices do not need to be raised

- Ensure all dietary requirements have been taken care of

- Match drinks to courses and explain to guests what they are eating and drinking

- Make sure you have a selection of teas and coffees for those who prefer something a bit different.

GAMES

Can you eat a doughnut without licking your lips? And how's your poker face? Here you will learn how to play it cool so you don't give the game away to your opponents, as well as how to master a host of new games and activities.

TURN RIGHT

For this activity, all you need is yourself. The rules are simple: go outside and start walking. Each time you reach a junction you can only keep going forward or turn right. It's deceptively simple – this is a great way to discover something new. You may even find a hidden gem of a place right round the corner! Try this in a car or on a bike if you want to travel further.

KEEP A POKER FACE

The key to a good poker game is keeping your hand to yourself. It's human nature to show emotion on your face when you get a good or bad hand, so the first rule is to put a stop to that. First you need to concentrate on relaxing your face and be mindful to keep it relaxed. Make sure you maintain good eye contact with the other players and blink occasionally – don't stare. Relax your posture, keep your words to a minimum and reply calmly when spoken to. Good luck!

EAT A DOUGHNUT FROM A STRING

This is a good one for Halloween (or any time you want to have a little fun) and is a real test for your self-control. Sugared ring doughnuts are best for this game as they are easy to attach. Simply thread each doughnut onto the string, then tie the string between two points, making sure it's high enough that the dangling doughnuts will be roughly at mouth height. Then let the players battle it out. You may not, under any circumstances, lick your lips until every bit of the doughnut has been swallowed.

PLAY BANNED WORDS

This game works best over the course of a meal, evening or weekend. Choose a word or phrase that is off limits for the duration of the time you will spend with the players. This could be anything from 'bathroom' or 'kitchen' to someone's name or an event or place. Whoever drops the banned word over the course of the game needs to then perform a forfeit.

THROW A BOOMERANG

When we think of a boomerang we tend to think of it returning – and although a boomerang can perform this trick, it isn't how it was traditionally used (which was as a weapon for hunting). In order to throw a boomerang and have it return to you, you need to hold the boomerang vertically, gripping it at the end of the bottom 'wing'. The nose of the boomerang should be facing out away from you. Then take the boomerang behind you and let go of it as if you were throwing a baseball.

PLAY I SPY

Sometimes the oldest games are the best. Make I spy more interesting by setting a specific theme or letter for the whole game and jolly up a dull car journey.

LEARN SOME BOARD GAMES FACTS

- The longest game of Monopoly ever played lasted 70 days.

- The Monopoly mascot used to be called Rich Uncle Pennybags before it was changed to Mr Monopoly.

- Snakes and Ladders originated in India in the nineteenth century, with the snakes and ladders representing karma – destiny and desire, respectively.

- Cluedo, originally called 'Murder!', was patented in 1947 but wasn't produced until 1949 due to post-war shortages. Original weapons included a bomb, syringe and fireplace poker.

- The highest score ever recorded for a word in Scrabble was 392 with 'caziques' (tropical American birds) and the highest score for a game was 1,049.

  - Scrabble was originally called 'Lexico'.

- Jenga is Swahili for 'build' and the game has sold more than 50 million copies, which equates to over 2.7 billion bricks.

  - Trivial Pursuit was invented by two Canadians who found there were tiles missing from their Scrabble set and so made up their own game.

- Chess originated in India around AD 280–550 and the pawns, knights, bishops and rooks were originally infantry, cavalry, elephants and chariotry.

PLAY HOW MANY PEGS...?

Sort of like human Buckaroo, the object is to see how many clothes pegs you can attach to other players over the course of a day or evening without them realising. Whoever secures the most pegs wins the game. This is a good one to play alongside another game, so everyone is distracted. Buy pegs in a selection of sizes and assign a points system, so giant pegs score the highest and miniature pegs score less. Ask players to write their initials on their assigned pegs.

PLAY CONSEQUENCES

Each player needs a piece of paper and a pen. This game has six rounds and between each round the paper will be passed to the player on the left. Each player should first write the name of a person, fold over the paper enough to cover what they have written and pass it on. Then should come the name of another person, a place (where they met), then what person 1 says to person 2, then what person 2 says back, then the consequence (what happened next). Each time something is written the paper is folded over enough to cover it. Pass it on a final time, unfold and read out to much hilarity.

PLAY PICTURE CONSEQUENCES

Try the above, but with pictures instead and three rounds – one each for the head, the middle and the bottom half – nothing is off limits, so animals/monsters/humans are all welcomed.

PLAY WINK MURDER

This is best played around a dinner table or at a party when people are distracted by conversation. Assign the murderer by placing as many pieces of paper in a hat as there are players, including one with the word 'murderer' on it. The murderer should then wink at other players to kill them off. Players that are still alive can accuse if they suspect. A victorious murderer will most likely kill all players but one!

CONCLUSION

As you've made it this far, I hope the preceding pages have demonstrated just how limitless the options are when it comes to reducing your screen time and doing something more imaginative, stimulating and hands-on! As incredible as modern technology has become, it's important to remember that there is a balance to be struck between being entertained by gadgets and doing things that can get us moving and thinking in different ways. Try learning something new every day and you'll be surprised at what you can achieve.

IMAGE CREDITS

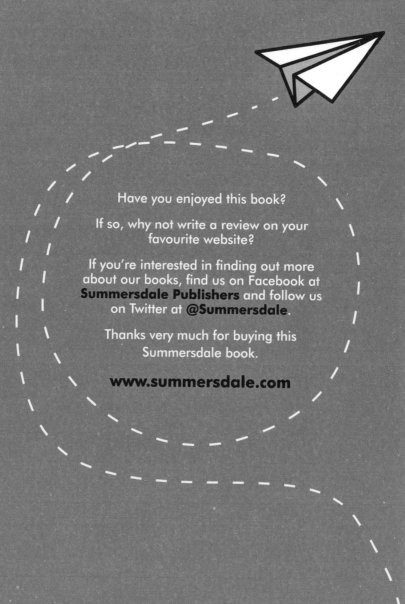

Have you enjoyed this book?

If so, why not write a review on your favourite website?

If you're interested in finding out more about our books, find us on Facebook at **Summersdale Publishers** and follow us on Twitter at **@Summersdale**.

Thanks very much for buying this Summersdale book.

www.summersdale.com